DOMINIE
ANIMAL LIFESTYLES

Fliers & Gliders

By Alison Ballance

Table of Contents

Dominie Press, Inc.

2

Introduction

People do not have wings. If we want to fly, we have to use machines. Airplanes have wings and engines. Hang gliders and paragliders use air **currents** and the wind to **soar** high in the sky.

Many animals fly and glide. They have wings or flaps of skin that let them soar through the air or **hover** above the ground. Animals that fly and glide can move faster through the air than they can on the ground. In this book, we will look at some of the animals that fly and glide.

Godwits

Godwits are long-distance fliers. Every year they fly from the far south of the world to the far north and back again.

Birds are perfect flying machines. Bird bones are light because they are **hollow**, but they are still very strong. Feathers are also very light, but they keep the bird warm.

The bar-tailed godwit flies 12,000 miles every year—from Alaska to New Zealand for the winter, and then back for the summer. It does not stop once along the way!

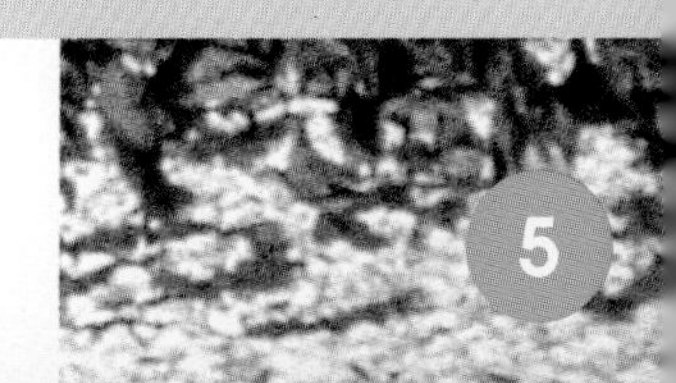

Albatrosses

Albatrosses live at sea and soar across the tops of waves. They do not flap their long, stiff wings very often. Instead, they catch wind currents and glide like a paper plane. Albatrosses can soar a long way without using much **energy**. They use the wind to help them stay in the air.

Albatrosses can stay in the air for a very long time. They are excellent fliers, but they often stumble when they take off and land.

Hummingbirds

Hummingbirds are the only birds in the world that can fly backward. They are very small birds. They flap their wings so fast they look like a **blur**. By beating their wings quickly, hummingbirds can hover in front of a flower while they feed on its nectar.

Because their wings move very fast, hummingbirds must visit hundreds of flowers in a day in order to have enough energy to keep flying.

Bumblebees

It is difficult to imagine how bumblebees can fly. They are heavy insects with small wings. But bumblebees are very good fliers. They make up for being heavy by having strong **muscles** to flap their wings.

Bumblebees produce honey, like other bees. But they do not produce very much of it at a time. They are not a good source of honey for people.

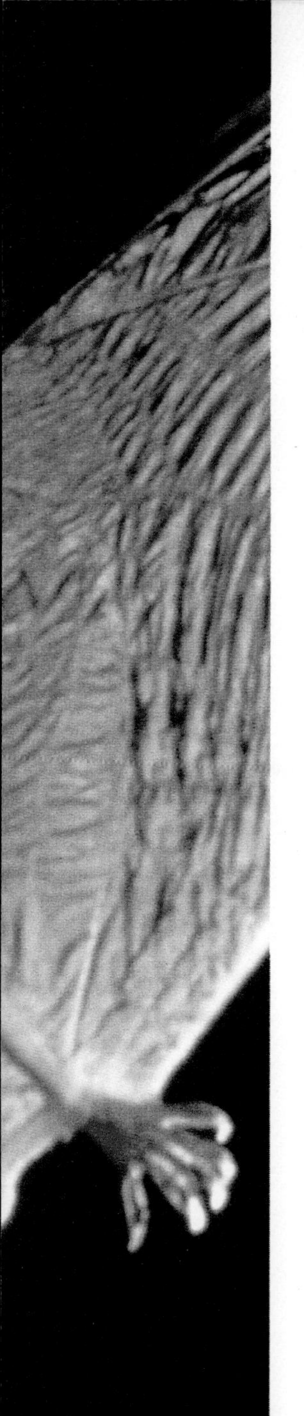

Bats

Bats are **mammals** that fly. They have big wings like birds, but they do not have feathers. Instead, they are covered in fur. They are quick fliers. They **swoop** around trees in the forest, catching insects in midair.

A bat's wings are actually hands that have thin flaps of skin between the fingers. When a bat is on the ground, it walks on its thumbs!

Flying Squirrels

A flying squirrel can jump out of a tree and glide for hundreds of feet to another tree. But a flying squirrel does not have wings. It has skin stretched between its front and back legs that acts like a sail on a ship. A flying squirrel changes **direction** by moving its flat tail.

*Flying squirrels, unlike other squirrels, are **nocturnal**—they are active only at night.*

16

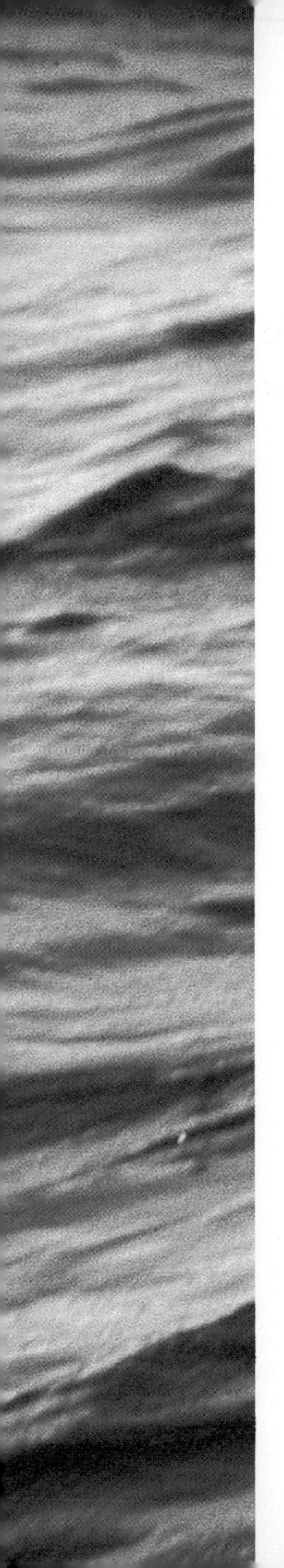

Flying Fish

Flying fish do not really fly—they **glide**. They leap out of the water when they are being chased. Their big fins act as wings. Flying fish can glide through the air for hundreds of feet. Sometimes they jump so high, they land on the decks of ships.

The flying fish uses its tail like an engine. It flips its tail from side to side very fast, like a hummingbird's wings, in order to get enough speed to launch itself into the air.

Butterflies

Butterflies have little bodies and big wings. They look **delicate**, but they are strong fliers. Monarch butterflies can fly for hundreds of miles.

Monarchs spend the winter in the southern United States and Mexico. In spring, they fly north to feed and breed. In fall, the next generation of butterflies flies south again.

*After **emerging** from its **chrysalis**, it takes an hour or more for a butterfly's wings to dry and take shape.*

Peregrine Falcons

The peregrine falcon is the fastest bird in the world. It dives from high in the sky to catch smaller birds. When peregrine falcons dive, they tuck their wings in to help them go faster. They have **powerful** claws to catch their prey and hold on while flying fast through the air.

When they are flying straight down, peregrine falcons can reach speeds of more than 200 miles per hour.

Summary

Some animals, like the butterfly, fly because they are looking for food. Other animals, like flying fish, leap into the air to escape predators who can't fly. They can move faster through the air than they can through the water.

When an animal flies, it must take care not to use too much energy. The wings of an albatross are very big. They take a lot of energy to flap, so the albatross uses the wind to glide, instead.

Because people can't fly, we use airplanes when we want to go somewhere fast. Airplanes carry a lot of fuel, which is energy for the engines. They can fly a long way without stopping.

Glossary

blur something that you can't see clearly

chrysalis a protective covering, made of silk by caterpillars

currents moving air

delicate fine and easily broken

direction the course something takes when it moves

emerging coming out of something

energy the power a body uses to move and do things

glide to move smoothly, without making an effort

hollow empty in the middle

hover to hold a position in the air

larvae the first stage of life for an insect (singular—larva)

mammals animals that have warm blood and suckle their young

muscles the parts of an animal's body that help it move

nocturnal active at night

powerful very strong

soar to fly without having to flap wings

swoop to fly up or down very quickly

Index